Our Not a Star!

by Peter Martinek
illustrated by Linda Helton

Copyright © by Harcourt, Inc.

All rights reserved. No part of this publication may be reproduced or transmitted in any form or by any means, electronic or mechanical, including photocopy, recording, or any information storage and retrieval system, without permission in writing from the publisher.

Requests for permission to make copies of any part of the work should be addressed to School Permissions and Copyrights, Harcourt, Inc., 6277 Sea Harbor Drive, Orlando, Florida 32887-6777. Fax: 407-345-2418.

HARCOURT and the Harcourt Logo are trademarks of Harcourt, Inc., registered in the United States of America and/or other jurisdictions.

Printed in China

ISBN 10: 0-15-350510-9
ISBN 13: 978-0-15-350510-2

Ordering Options
ISBN 10: 0-15-350334-3 (Grade 4 Below-Level Collection)
ISBN 13: 978-0-15-350334-4 (Grade 4 Below-Level Collection)
ISBN 10: 0-15-357499-2 (package of 5)
ISBN 13: 978-0-15-357499-3 (package of 5)

If you have received these materials as examination copies free of charge, Harcourt School Publishers retains title to the materials and they may not be resold. Resale of examination copies is strictly prohibited and is illegal.

Possession of this publication in print format does not entitle users to convert this publication, or any portion of it, into electronic format.

We're going on vacation. We're driving to
   the hills.
We're bound to have some fun. We hope to have
   some thrills.

Dad says that Montana is a pretty place to see.
Mom tells us Montana is the place we want to be.

I heard that there are forests there and many
   sparkling lakes.
I'm just a little nervous, though. I don't like bears
   or snakes.

Of course, the only problem is we have to drive
   our car.
Mom's a little worried because our car is not a star.

You see, our car is rather old and doesn't like to go.
Sometimes when we turn the key, the car seems
   to just say "no."

Mom says we should take it in for some overdue
   repairs.
Dad says not to worry. "It will surely get us there."

The car is filled with suitcases, tents, and many bags.
The trunk is overloaded, and the back end almost drags.

We all get inside with Mom and Dad in the front seat.
The rest of us get in back and start to feel the heat.

There is no air conditioning. It's very hot outside.
Our bodies start to sizzle on our bumpy, awful ride.

The seats are old and scratchy, and our legs start to get sore.
When Melissa stretches out, she crushes me against the door.

Gary starts to push my arm, and then I start to cry.
Melissa just ignores me and keeps staring at the sky.

I think how I am missing playing games with all my friends.
I wish that this vacation would come quickly to an end.

After several hours of driving, we very quickly stop.
Of course, one of our tires just had to go and pop.

Dad takes a while to fix it while we sit in the
   hot sun.
Our old and ugly car is ruining our vacation fun.

Now we have three tires and a patched-up
   wobbly spare.
Since we are now back on the
   road, we really do not care.

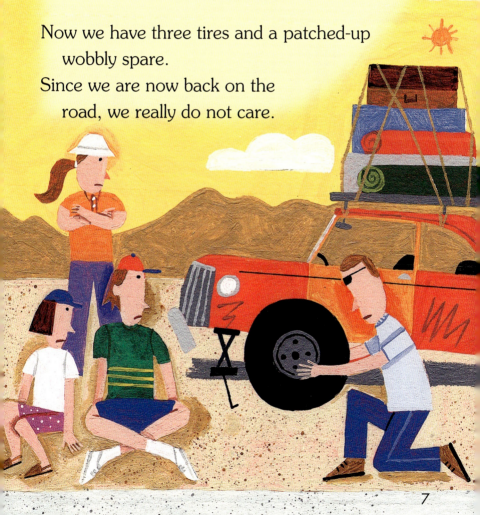

As the light begins to fade, we drive to a motel.
The thought of leaving that old car makes us all
 feel swell.

When Dad finally parks the car, Gary jumps out
 first.
I'm glad to leave that cramped back seat; it really
 is the worst.

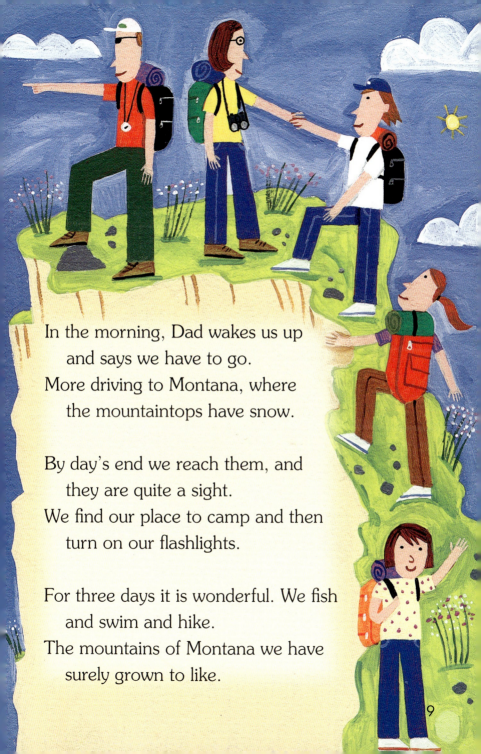

In the morning, Dad wakes us up
and says we have to go.
More driving to Montana, where
the mountaintops have snow.

By day's end we reach them, and
they are quite a sight.
We find our place to camp and then
turn on our flashlights.

For three days it is wonderful. We fish
and swim and hike.
The mountains of Montana we have
surely grown to like.

Then the dreaded day comes when we have to drive on back.
No more strolling through the woods while carrying a backpack.

Now it's the time to get ourselves into that back seat.
With no choice we surrender to the bumps and mess and heat.

We are clustered all together like sardines in a tin.
Dad starts up the old engine, and our long ride back begins.

We drive for several hours, then we hear an awful
   sound.
In another moment, we can feel the car slow down.

Smoke pours from the engine as we scramble out
   the door.
It soon becomes apparent that the car will run
   no more.

We gather up our suitcases, and we walk through
   pouring rain
To reach a nearby town where we can catch an
   eastbound train.

"What about the car, Dad?" we ask as we sit down.
"The man says he can't fix it," Dad tells us with
   a frown.

Sitting on this nice cool train, it really is a treat.
We all have room to spread our elbows, knees,
    and tired feet.

The ride is smooth and easy as we roll along
    for miles.
The train is particularly comfortable, and
    everybody smiles.

Mom stares out the window at the mountains,
    lakes, and trees.
Dad looks at us all and smiles because we are
    at ease.

Dad says, "There's a dining car where we can go to eat."
As soon as we walk in, we say, "This place is really neat."

A half an hour later, they bring chicken, corn, and beets.
After we have eaten, they bring out a tray of treats.

As we pull into the station and get off our
   fancy train,
Gary says, "I hope that we can soon ride the
   rails again."

Dad says we will take a cab from the station to
   our place.
We stand there on the sidewalk holding many
   a suitcase.

Then we grab the bags and cram into the cab's
   back seat.
"Here we go again," I say. "No more room for
   knees or feet!"

Think Critically

1. Why did the family need to ride a train?

2. How is the train ride different than the car ride?

3. What is different about the way this story is written? Explain your answer.

4. Does anyone in the story remind you of someone you know? If so, how?

5. What part of this story did you think was the funniest? Why?

☆ Language Arts

Write a New Ending Think of another way the family could have gotten home. Then write a different ending for this story. If you wish, you may try to make it rhyme like the poem. If not, write a paragraph telling your new ending.

 School-Home Connection Tell friends or family members about this poem. Then talk about problems that can come up when you are away from home and what you might do about them.

Word Count: 797